FLORA OF TROPICAL EAST AFRICA

CAPRIFOLIACEAE

B. Verdcourt

Small trees, shrubs, woody climbers or rarely herbs, often with soft pith. Leaves opposite or less often alternate*, simple or pinnately compound; stipules absent or very small, rarely conspicuous. Inflorescences mainly cymose, less often flowers solitary or paired; bracts and bracteoles usually present. Flowers regular or irregular, hermaphrodite. Calyx-tube joined with the ovary; limb (3–)4–5-fid or -toothed, the teeth imbricate or open. Corolla gamopetalous, tubular, funnel-shaped, campanulate or rotate, 3–5-lobed, sometimes 2-lipped, the lobes imbricate or less often valvate. Stamens 4–5, inserted in the corolla-tube and alternate with the lobes; anthers 2-thecous, opening lengthwise inwards or outwards. Ovary inferior, 2–8-locular, with 1-many pendulous ovules in each locule; placentation axile or parietal; style simple or wanting; stigma 1, capitate, or stigmas as many as the carpels. Fruit a berry or drupe, less often a capsule or achene, 1–5-locular with 1–many seeds or as many pyrenes as locules; endosperm copious; embryo mostly small and straight.

A widespread family of 13–18 genera (according to treatment), represented in tropical Africa by only the single species treated below.

The family contains numerous species cultivated as ornamentals, mostly in the genera *Viburnum* L., *Weigela* Thunb. and *Lonicera* L. Jex Blake, Gardening in East Africa, ed. 4 (1957), mentions several species, but it is not clear whether they have been cultivated or would be suitable for cultivation. It appears that in addition to *Abelia* × *grandiflora* (André) Rehd., mentioned below, *A. floribunda* (Mart. & Gal.) Decne. and *Viburnum tinus* L. (the common laurustinus) have been cultivated in the highlands of Kenya. Herbarium material has been seen of several species, namely *Abelia* × *grandiflora* (Nairobi, Karen, *Verdcourt* 1775!); *Lonicera* × *americana* (Mill.) Koch (Nairobi, *Grahame Bell* in *E.A.H.* 10844! & 10845!); *L. hildebrandiana* Coll. & Hemsl., a very long-tubed honeysuckle (e.g. Kenya, N. Nyeri District, Nanyuki, Mawingo Hotel, *Conway Evans*!, Kiambu District, Muguga, Hort. Greenway, *Greenway* 10870! and Nairobi Arboretum, *G. R. Williams* 452!, also Tanganyika, Amani, *Mrs. Moreau* in *Herb. Amani* 8696!); *L. japonica* Thunb., mostly var. *halliana* (Dipp.) Nichols. (e.g. Nairobi, *Jex Blake* in *E.A.H.* 10864! & H. 29/56 & *Grahame Bell* 11!) and *Sambucus canadensis* L. (see below).

SAMBUCUS

L., Sp. Pl.: 269 (1753) & Gen. Pl., ed. 5: 130 (1754)

Small trees, shrubs or rarely perennial subshrubby herbs, mostly evil-smelling. Leaves pinnately or bipinnately compound; leaflets usually serrate; stipules absent, small or occasionally conspicuous, sometimes reduced to clusters of glands. Inflorescences usually terminal, corymbose, cymose or thyrsoid. Flowers small, regular. Corolla white or yellowish, rotate, the lobes imbricate or valvate. Stamens 5; filaments slender; anthers dehiscing outwards. Ovary 3–5-locular, with a single pendulous ovule in each locule; style wanting, the stigma sessile and 3–5-lobed. Fruit

* In those genera referred by Airy Shaw to *Alseuosmiaceae*.

1

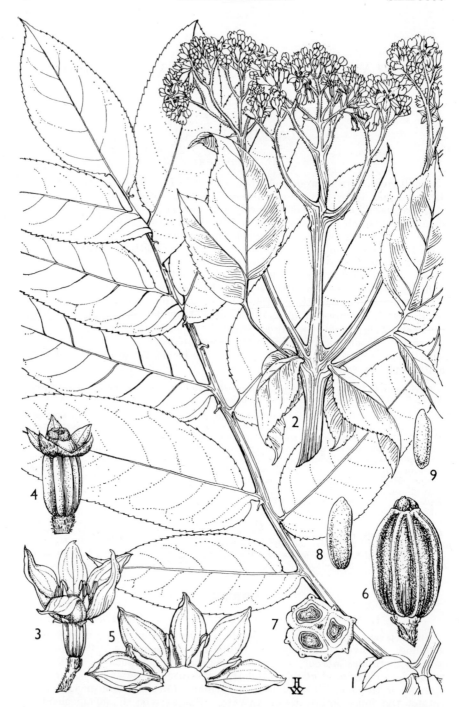

FIG. 1. *SAMBUCUS AFRICANA*—**1,** leaf and small portion of branch, × ½; **2,** part of flowering branch, × ½; **3,** flower, × 3; **4,** same with corolla and stamens removed, × 5; **5,** corolla and stamens, × 3; **6,** fruit, × 4; **7,** transverse section of fruit, × 4; **8,** pyrene, × 4; **9,** seed, × 4. 1, from *Fries* 681; 2, from *Armstrong* in *Napier* 2085; 3–9, from *Greenway* 9703.

a berry-like drupe with 3–5 1-seeded pyrenes. Seeds with a membranous testa and fleshy endosperm.

A genus of about 20–30 species according to delimitation occurring chiefly in the northern temperate regions but extending to East Africa, Central and South America, Indochina, India, Malesia, Philippines, Australia and New Zealand.

Some workers have placed *Sambucus* in a separate family *Sambucaceae*, showing affinities with the *Staphyleaceae* and *Valerianaceae* and differing in a number of important characters from the rest of the *Caprifoliaceae*; the genus *Viburnum* is, however, somewhat intermediate and I am not at all convinced that the segregation is justifiable.

A single species occurs in East Africa, allied to *S. ebulus* L. (Europe, N. Africa) and *S. adnata* DC. (Himalayan region). *S. canadensis* L. has been cultivated (e.g. Kenya, Uasin Gishu District, Eldoret, Oldane Sapuk, *Greenway* 8514!; Nairobi, Closeburn Nursery, *Grahame Bell* in E.A.H. 10449! & *Greenway* 8778!; Tanganyika, Mbulu Forest Nursery, *Matalu* in F.H. 3076!).

S. africana *Standl.* in Smithson. Misc. Coll. 68(5): 19 (1917). Type: Kenya, W. slopes of Mt. Kenya, *Mearns* 1746 (US, holo., BM, iso.!)

Shrubby herb, 1·2–4 m. tall, the roots spreading horizontally and producing stems at intervals. Stems mostly juicy but somewhat woody at the base, at first covered with multicellular hairs but later glabrous. Leaves imparipinnate, attaining 75 cm. in length and 45 cm. in breadth (*fide* A. S. Thomas), (5–)6–11-foliolate; leaflets subsessile, occasionally apical one partly fused with a lateral, elliptic-oblong, ovate or oblong-lanceolate, 4·5–18 cm. long, 1·3–8 cm. wide, acuminate, very unequally rounded at the base, finely and sharply toothed (the basal teeth often glandular), at first hairy, later glabrous above except for minute hairs on the main nerves, bullate, sparsely pubescent on the main nerves beneath which are slightly raised and reticulate; petioles 0·6–12 cm. long; stipules conspicuous, leafy, toothed (true nature uncertain); on the rhachis near the leaflet bases there are often glandular ? stipels. Inflorescences corymbose, 7–15 cm. wide, ferruginous pubescent, subtended by 1–3-foliolate glandular-toothed leaves; bracts and bracteoles linear, small. Calyx pinkish or purplish; tube ribbed; lobes triangular, ± 1 mm. long. Corolla white or creamy-white; tube 1·5 mm. long; lobes elliptic, 4–6 mm. long, 3–3·5 mm. wide, induplicate-valvate in bud. Fruit oblong or ellipsoid, 4·2–7 mm. long, 2·5–4 mm. wide, blue-black, 9–10-ribbed; fruiting pedicels purplish. Pyrenes 3–4, bony, pale reddish-brown, linear-oblong, 4–5·5 mm. long, 1–3 mm. wide, somewhat trigonous in section. Fig. 1.

UGANDA. Elgon, Butandiga, 14 July 1924, *Snowden* 908! & Elgon, Bugishu, near Sasa Hut, 22 Mar. 1951, *G. H. S. Wood* 102! & Benet Sabei, 12 Dec. 1938, *A. S. Thomas* 2637!
KENYA. NE. Elgon, July 1950, *Tweedie* 848!; Naivasha District: S. Kinangop, 2 Sept. 1951, *Verdcourt* 598!; Kiambu District: Kinale, 3 Aug. 1960, *Greenway* 9703!
TANGANYIKA. Masai District: Embagai Crater, Feb. 1954, *Eggeling* 6788! & 29 July 1962, *Newbould* 6222!
DISTR. **U3**; **K3**–6; **T2**; not known elsewhere
HAB. Upland grassland or evergreen bushland, clearings in or on the edges of moist bamboo thicket and upland rain-forest, or sometimes in upland rain-forest undergrowth; 1800–3150 m.

SYN. [*S. ebulus* sensu Engl., P.O.A. C: 374 (1895), *non* L.]
      S. ebulus L. var. *africanus* Engl. in Ann. Bot. 18: 537 (1904); T.S.K.: 149 (1936). Types: Kenya, Kikuyu, Abori, *Fischer* 327 (B, syn. †) & Kikuyu, *Elliott* 12 & 177 (both B, syn. †, K, isosyn.!)
      [*S. adnata* sensu Schwerin in Mitt. Deutsch. Dendrol. Ges. 1909: 41 (1909); Verdc. in K.B. 11: 445 (1957), *non* DC.]
      S. adnata DC. var. *puberula* Schwerin in Mitt. Deutsch. Dendrol. Ges. 1920: 221 (1920), pro parte, *nom. illegit.* based on typical *S. adnata*